AF680181

Eileen Chevalier's

TYPEWRITER Aalphabet

BOTANY Color Label

KNIGHT KING PRESS
Siler City, NC

written & illustrated by **Eileen Chevalier**

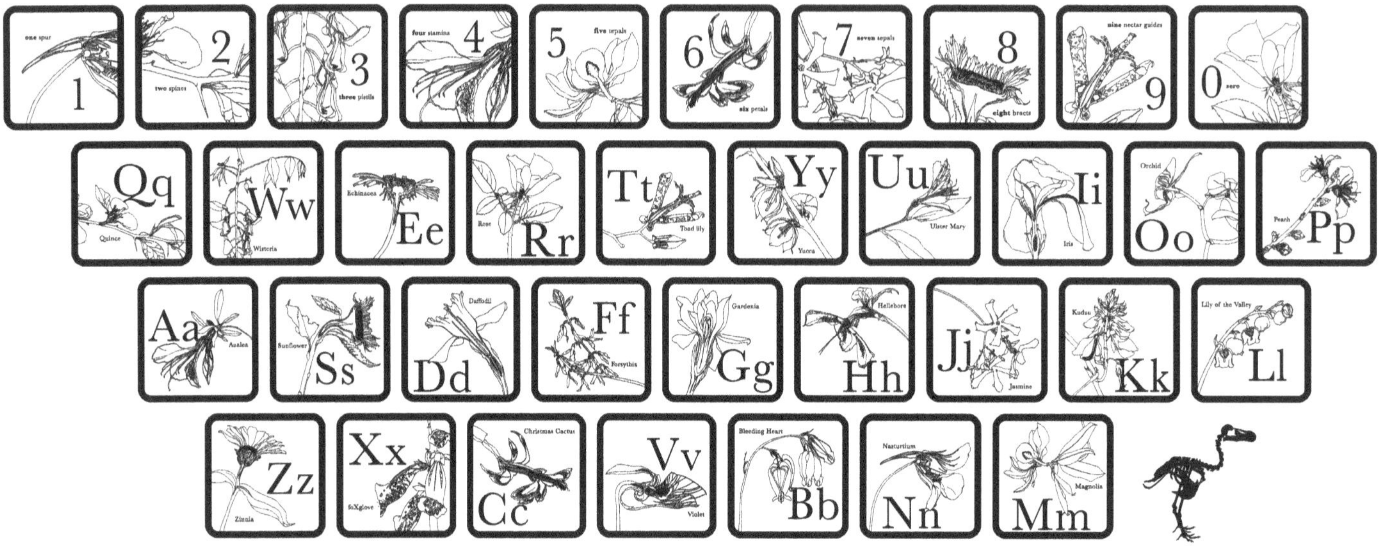

© 2026 Eileen Chevalier
Knight King Press

All rights reserved. No part of this publication may be reproduced, stored in a retrieval system, scanned or used in any form, or transmitted in any form or by any means (including by any form of Artificial Intelligence) without express written consent of the copyright holder and the publisher, Knight King Press, LLC.
ISBN 978-1-952704-14-7

Title: Eileen Chevalier's Typewriter Aalphabet Botany Color + Label Book / Eileen Chevalier; illustrated by Eileen Chevalier.
Names: Eileen Chevalier, author | Eileen Chevalier, illustrator. | Eileen Chevalier, designer.
Subjects: LCSH Botany | Plants–Anatomy–Juvenile literature. | Flowers–Pictoral works–Juvenile literature.
Classification: LCC QK49 .C48 2026 | DDC j580 C527 cl 2026

Also in the *Eileen Chevalier's Typewriter Aalphabet* Series:
Botany Book ISBN 978-1-952704-08-6 (hardback)
ISBN 978-1-952704-16-1 (paperback)
ISBN978-1-952704-17-8 (ebook)
Teacher's Guide ISBN 978-1-952704-13-0
Workbook ISBN 978-1-952704-15-4

The artwork in this book was hand drawn by Eileen Chevalier.
The text is set in Century Gothic, Avenir Next, and Baskerville.

www.knightkingpress.com

KNIGHT KING PRESS

Aa

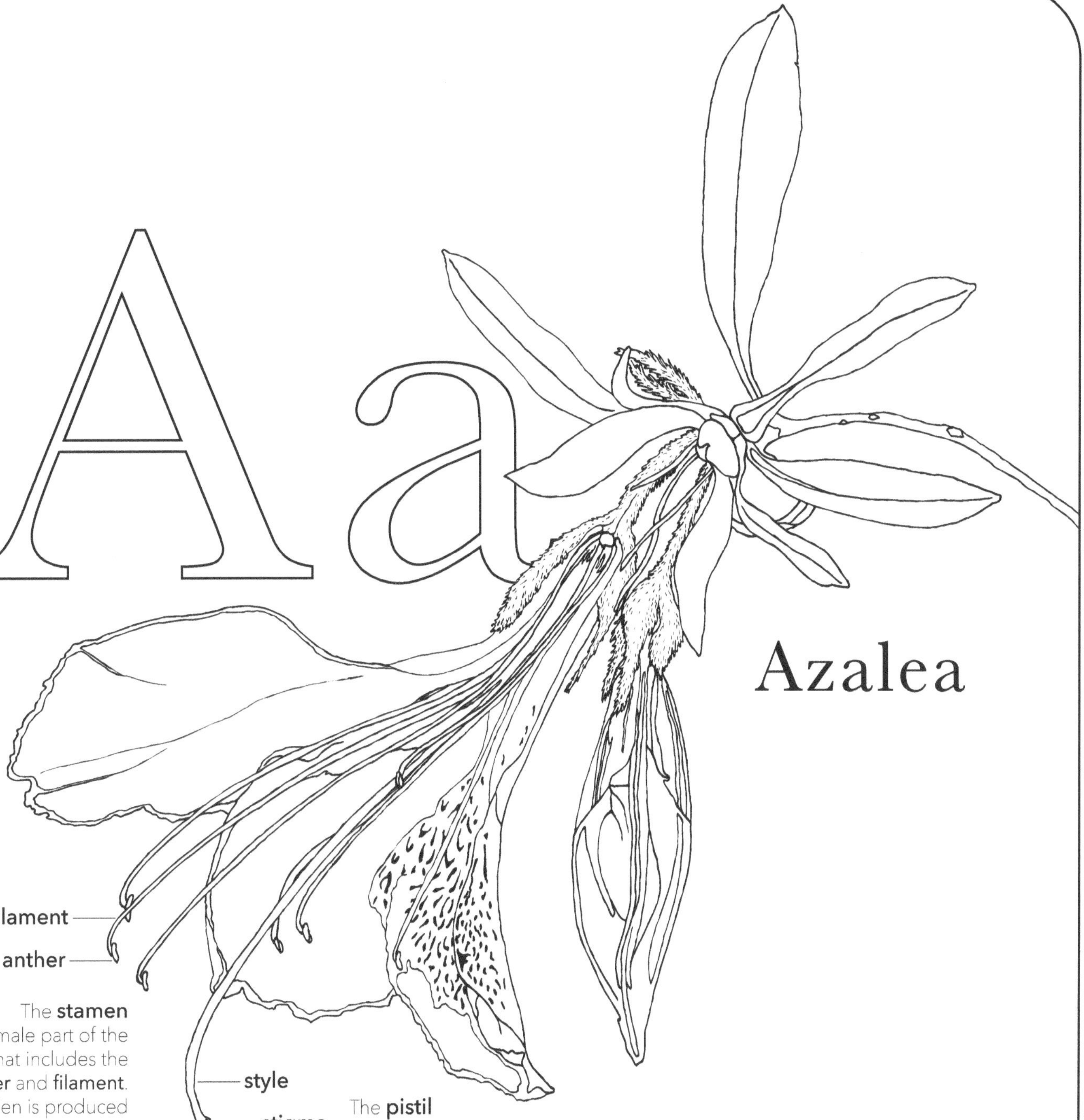

Azalea

The **stamen** is the male part of the flower that includes the **anther** and **filament**. Pollen is produced on the anther.

The **pistil** is the female part of the flower that includes the **stigma**, **style**, and **ovary**. The ovary is where the seeds are produced.

Bleeding Heart

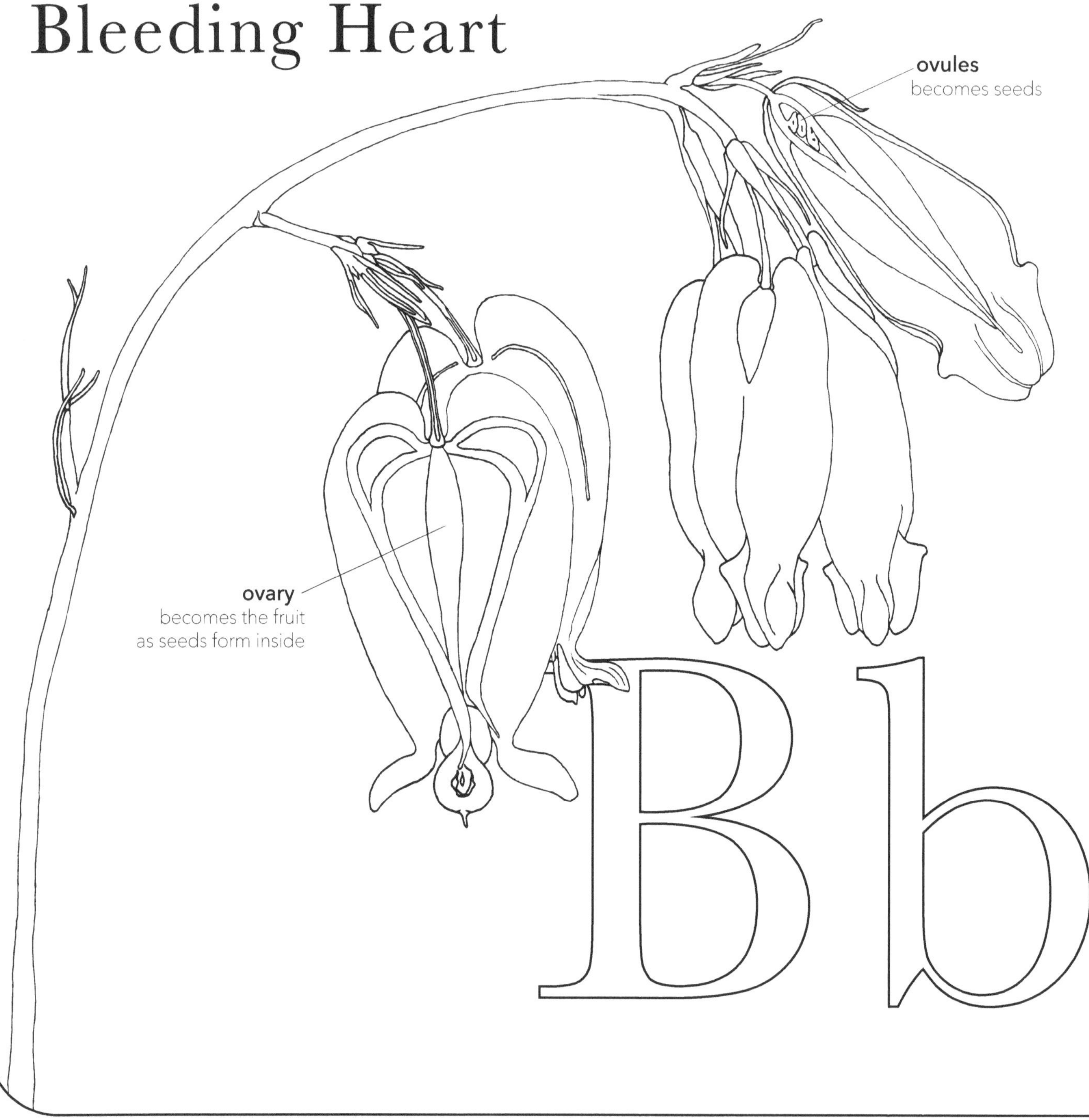

Christmas Cactus

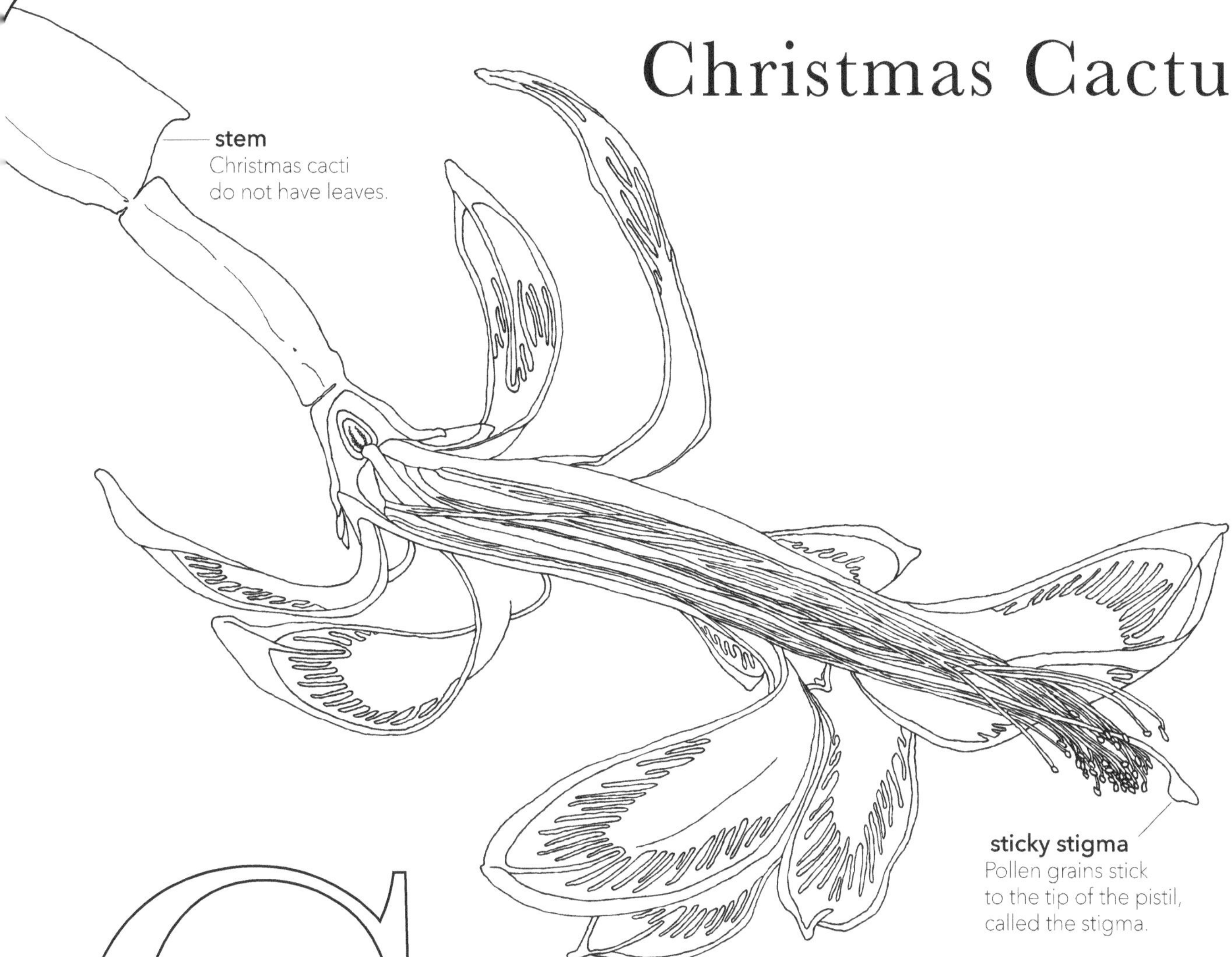

Cc

Daffodil

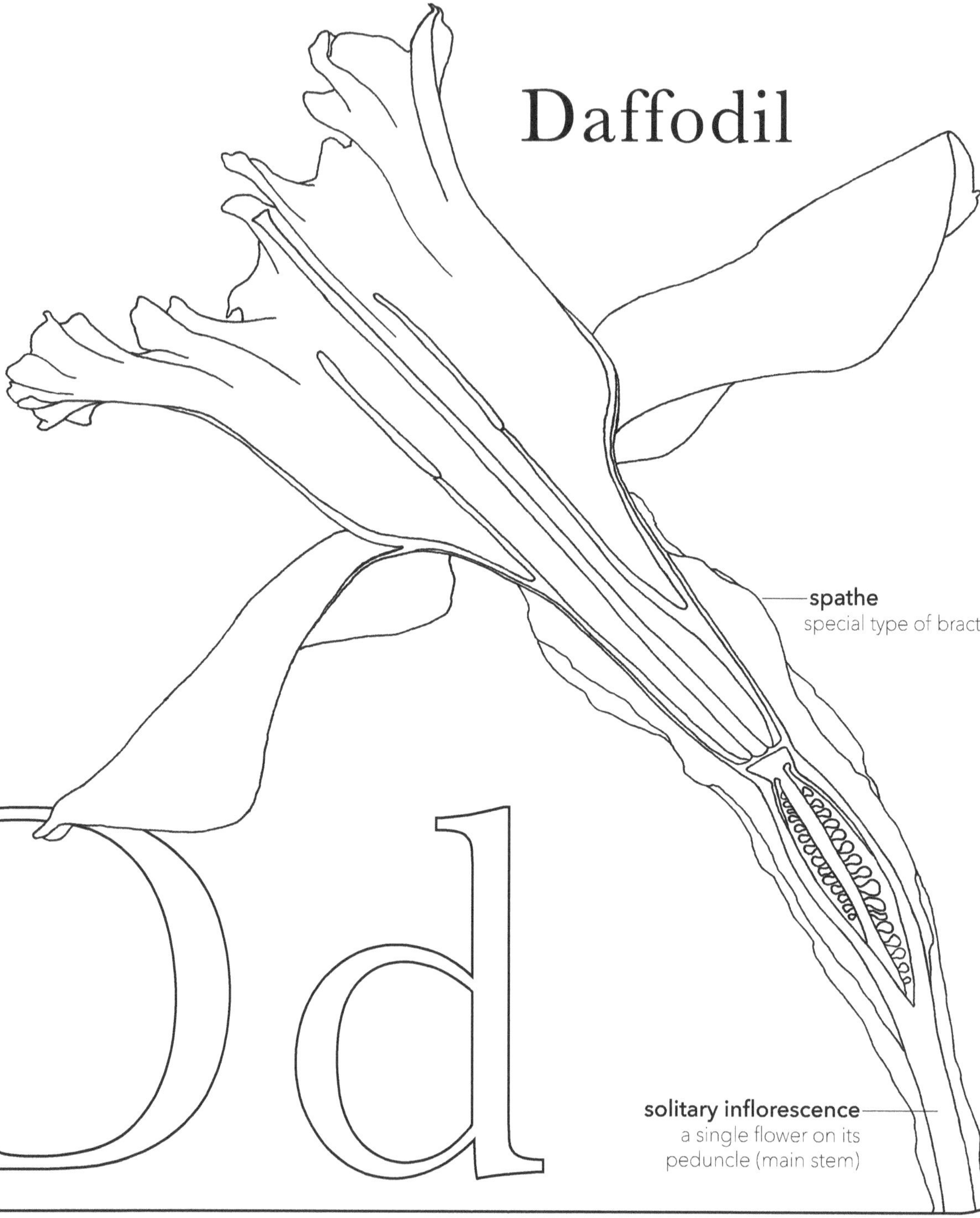
spathe
special type of bract
solitary inflorescence
a single flower on its
peduncle (main stem)

Dd

Echinacea

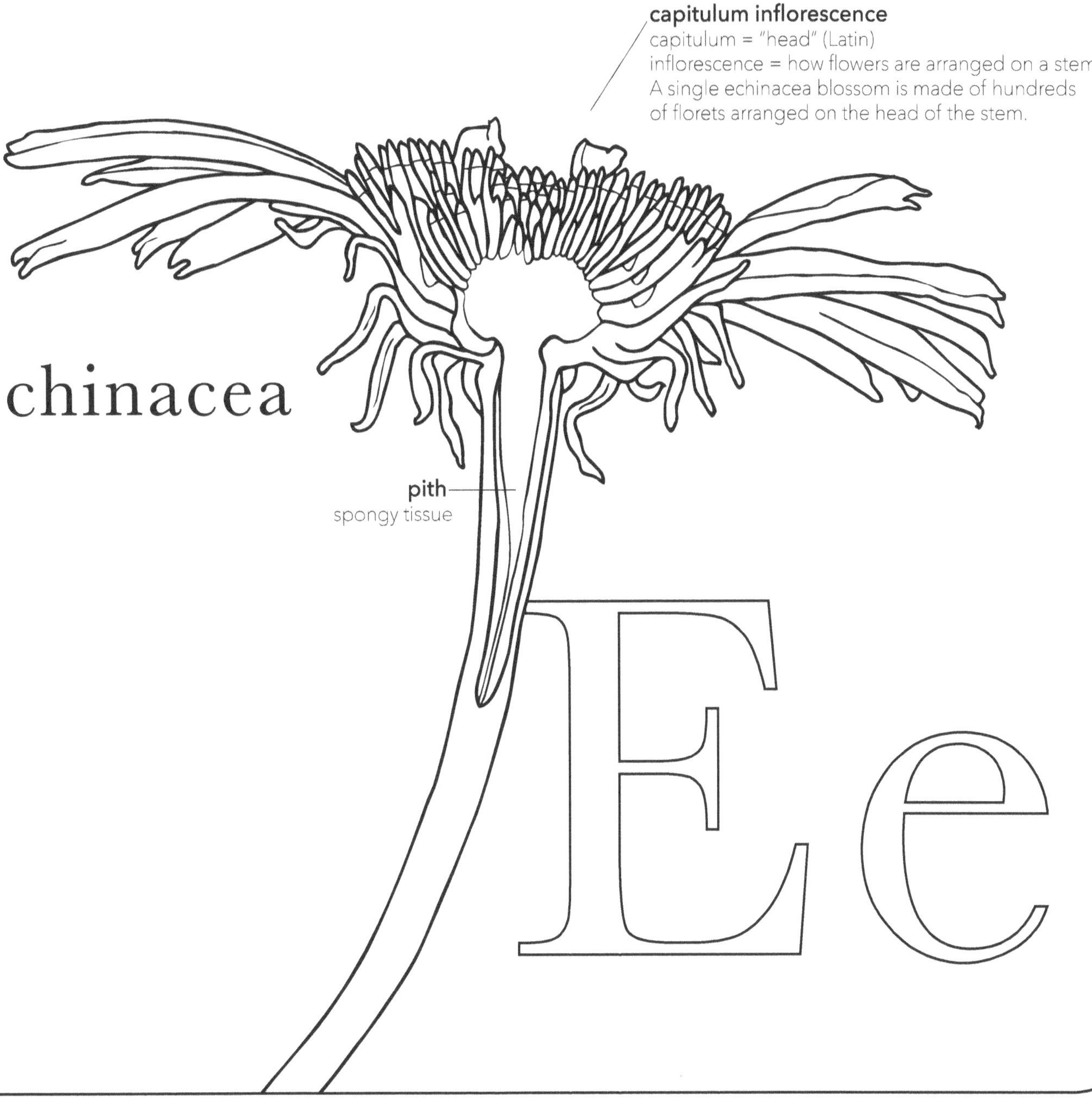
capitulum inflorescence
capitulum = "head" (Latin)
inflorescence = how flowers are arranged on a stem
A single echinacea blossom is made of hundreds
of florets arranged on the head of the stem.
pith
spongy tissue
Ee

Ff

Forsythia

new leaves
Forsythia loses its leaves every fall and produces new ones each spring. This means it is deciduous.

Gardenia

Gg

evergreen

Gardenia has leaves all year long.
This means it is evergreen.

Hellebore

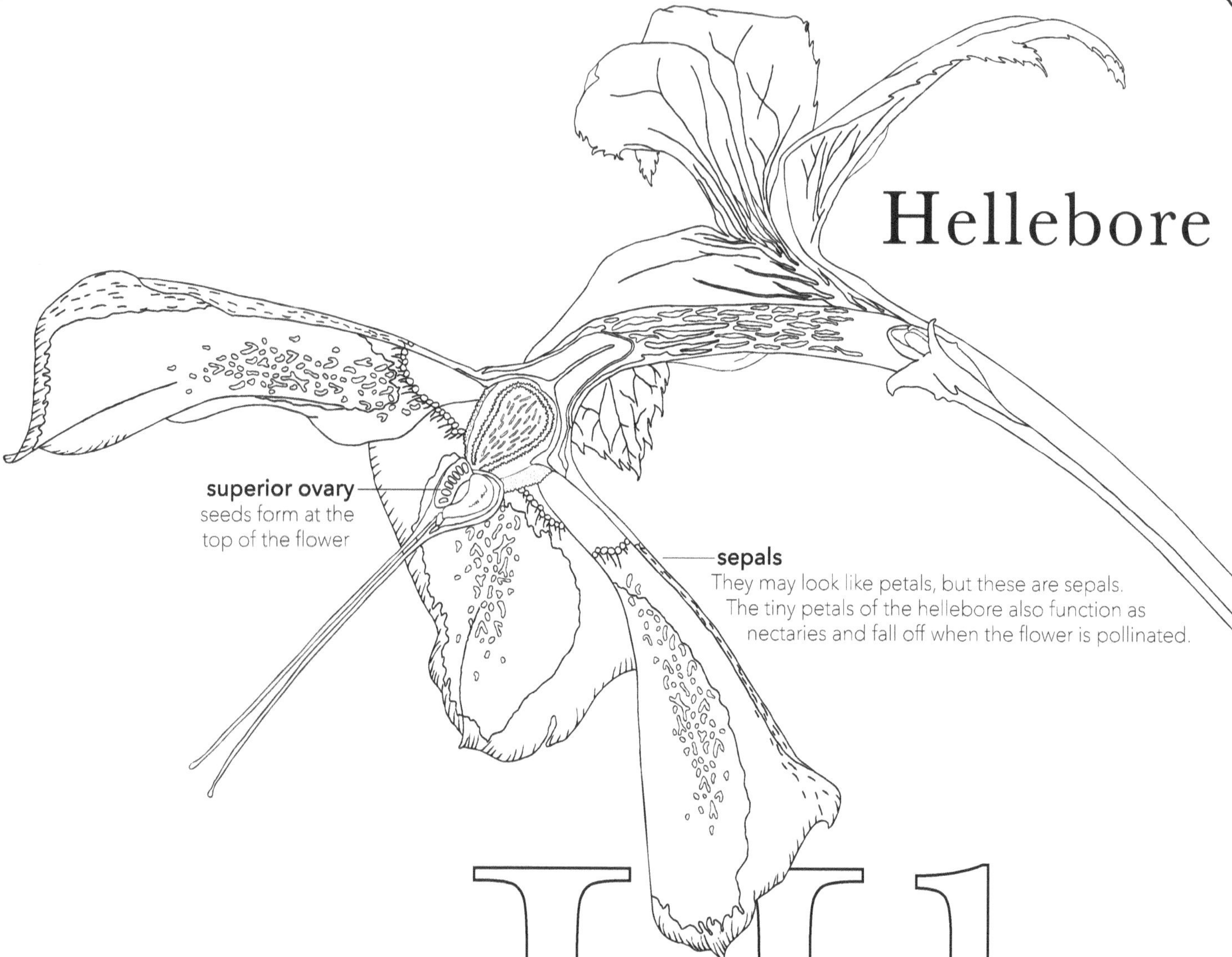

Hh

Ii

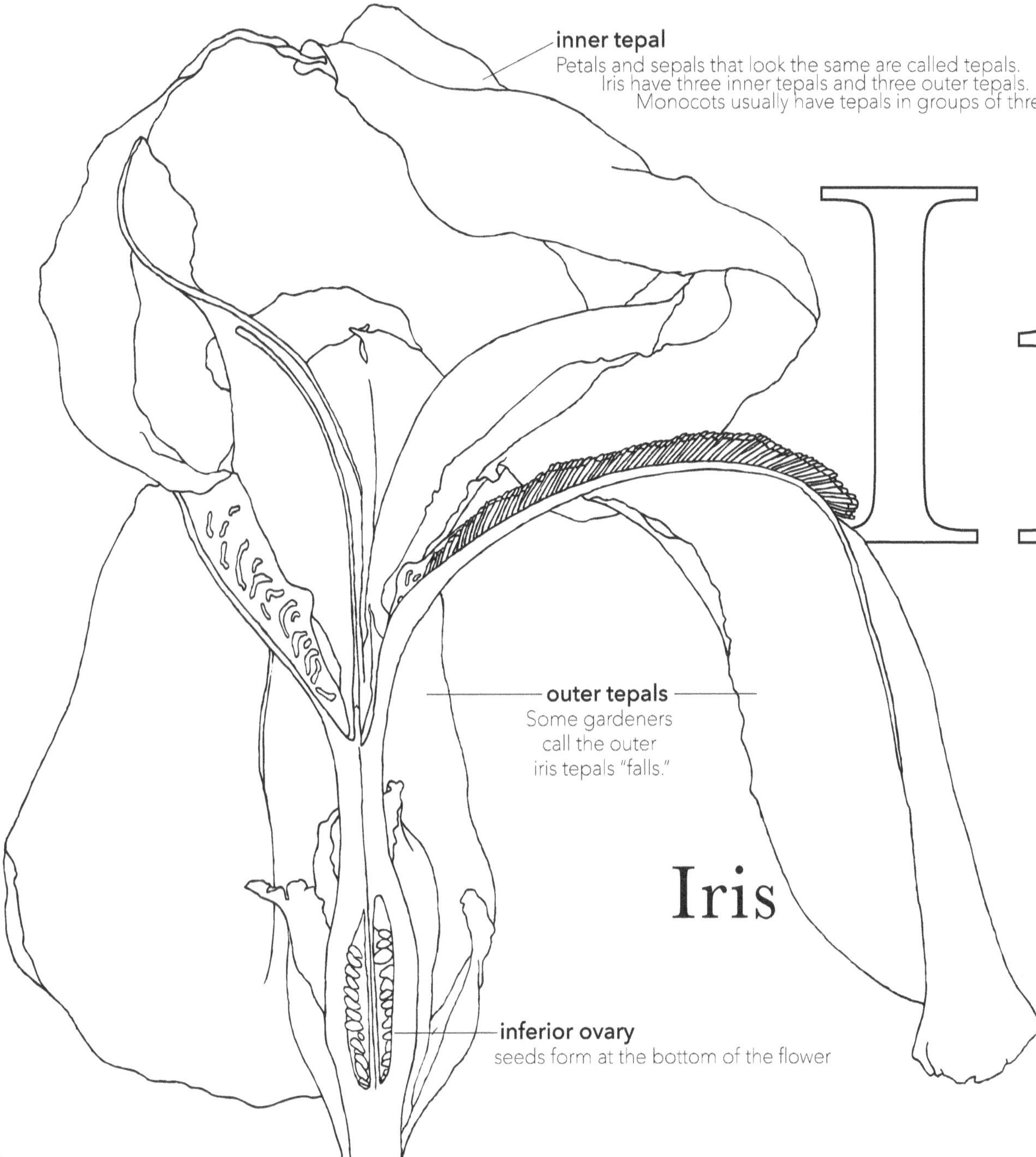

inner tepal
Petals and sepals that look the same are called tepals.
Iris have three inner tepals and three outer tepals.
Monocots usually have tepals in groups of three.
outer tepals
Some gardeners
call the outer
iris tepals "falls."
inferior ovary
seeds form at the bottom of the flower

Iris

Jj

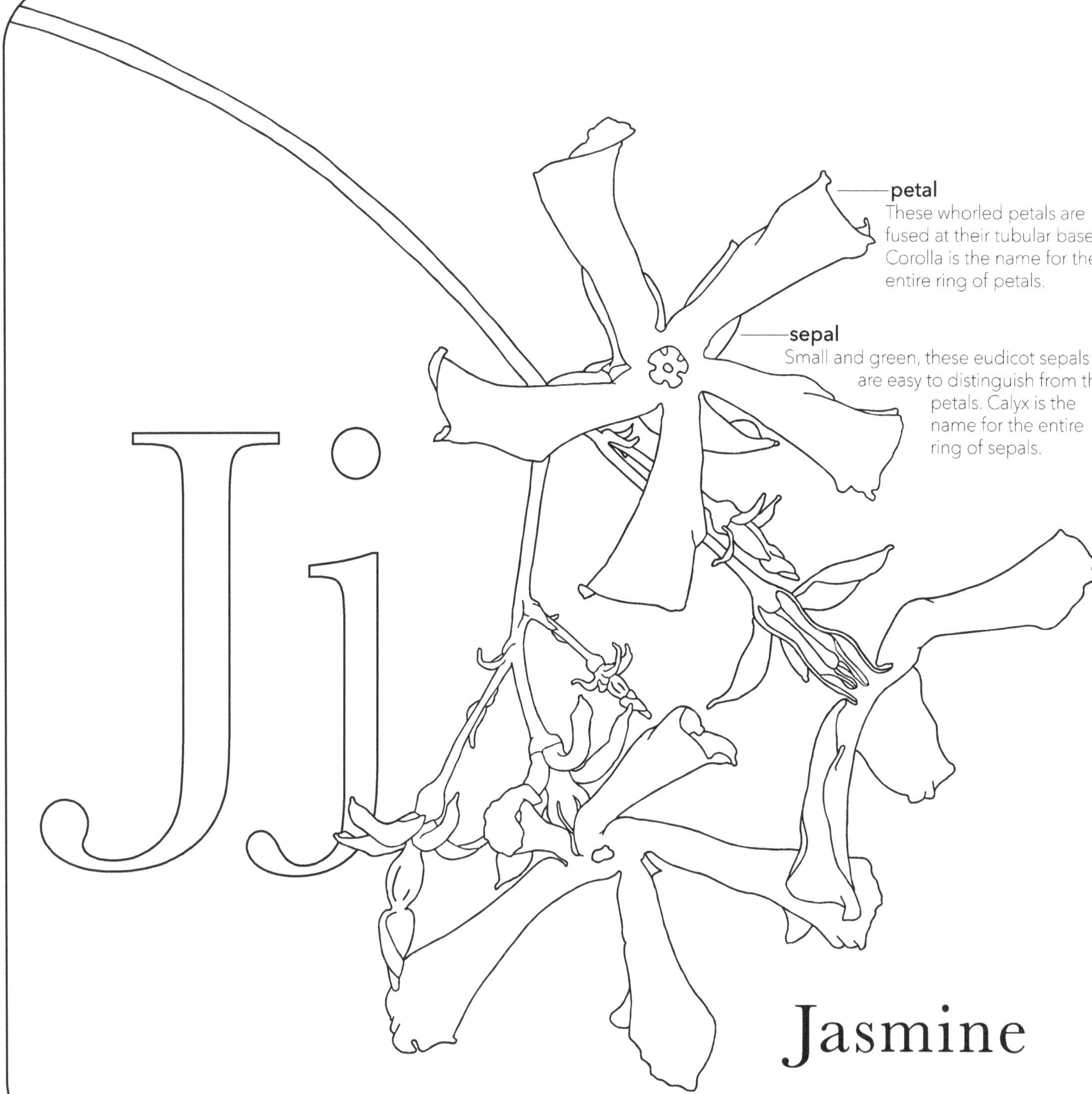
petal
These whorled petals are fused at their tubular base. Corolla is the name for the entire ring of petals.
sepal
Small and green, these eudicot sepals are easy to distinguish from the petals. Calyx is the name for the entire ring of sepals.

Jasmine

Kudzu

perianth
perianth = the calyx + the corolla
calyx = all the sepals of one flower
corolla = all the petals of one flower
Kk

Lily of the Valley

Ll

tepal
Magnolia is in the magnoliidae clade.
Magnoliids usually have tepals
like the monocot clade.

Magnolia

Mm

Nasturtium

Nn

Orchid

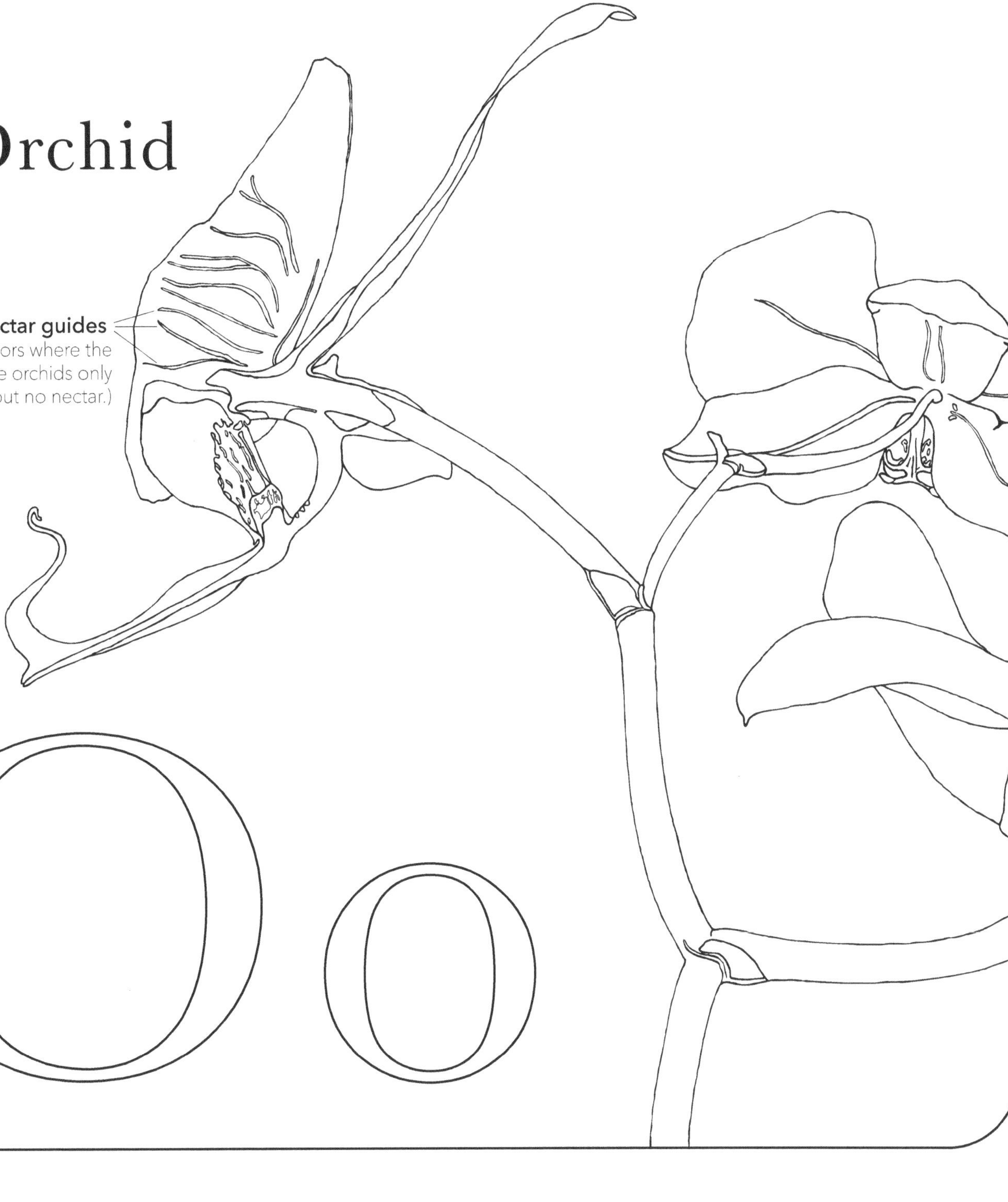
nectar guides
show pollinators where the nectar is (Some orchids only have guides but no nectar.)

Oo

fruit
All angiospermae (flowering plants) produce fruit. Some plants, like the dandelion, produce dry fruit and others, like this peach, produce succulent fruit.

Peach

Pp

Qq

serrated leaf margin
margin = edge of leaf
branched venation
venation = how veins are arranged in the leaf

Quince

Rose

prickles
Often mistaken for thorns,
prickles can appear anywhere on
the stem and are easier to remove.

Rr

Ss

Sunflower

Tt

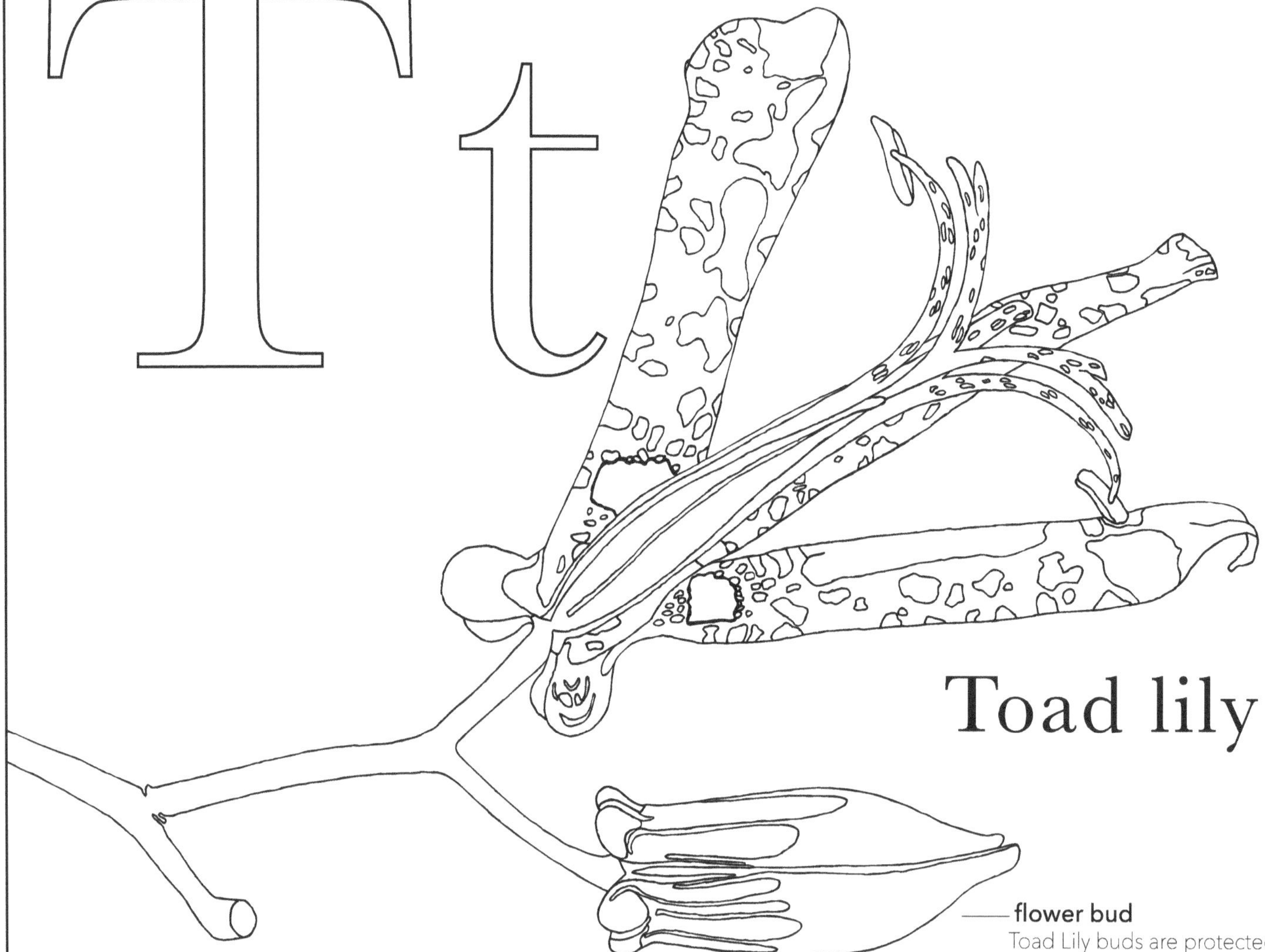

Toad lily

flower bud
Toad Lily buds are protected by their three outer tepals.

Uu

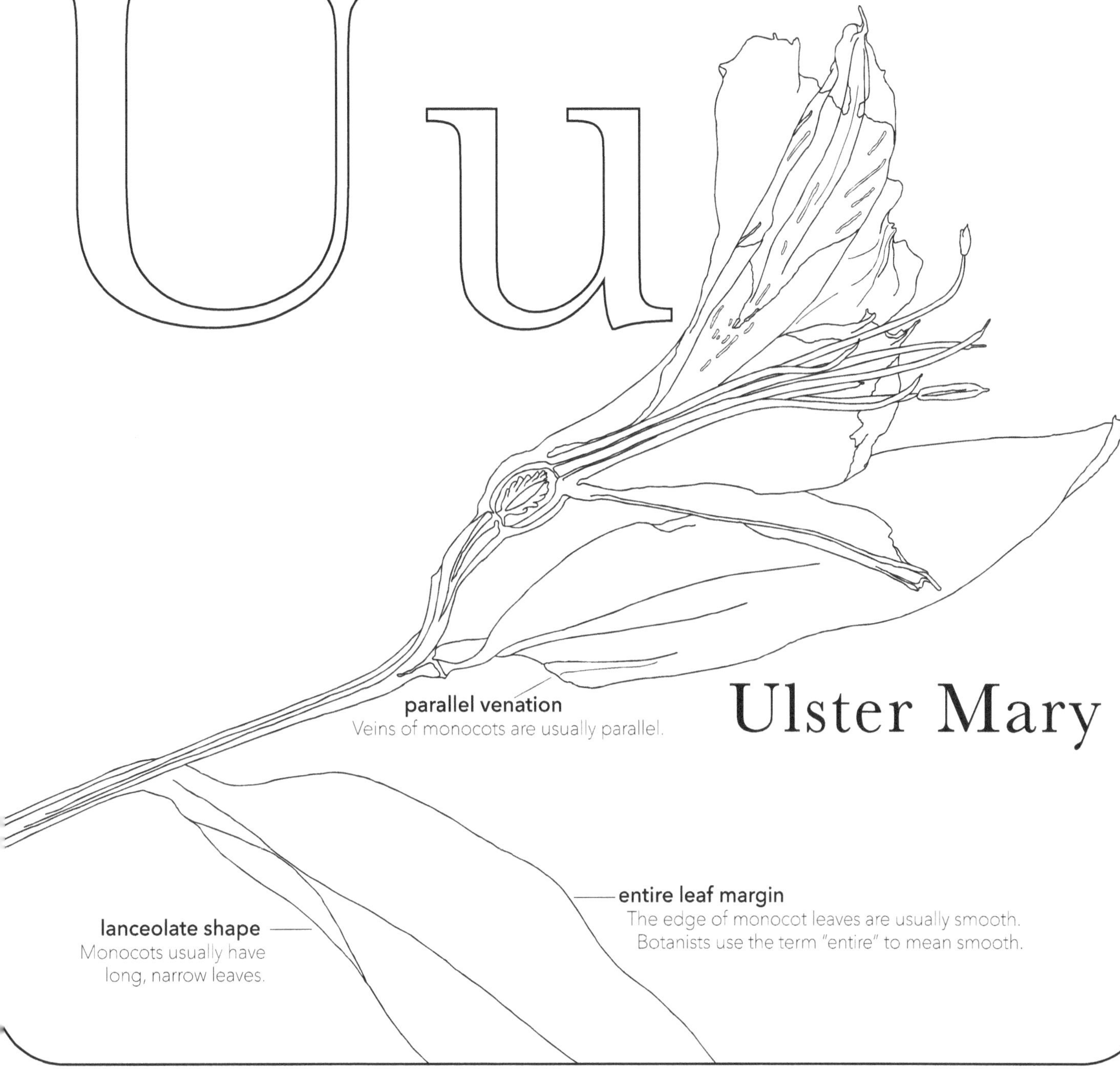

Ulster Mary

V v

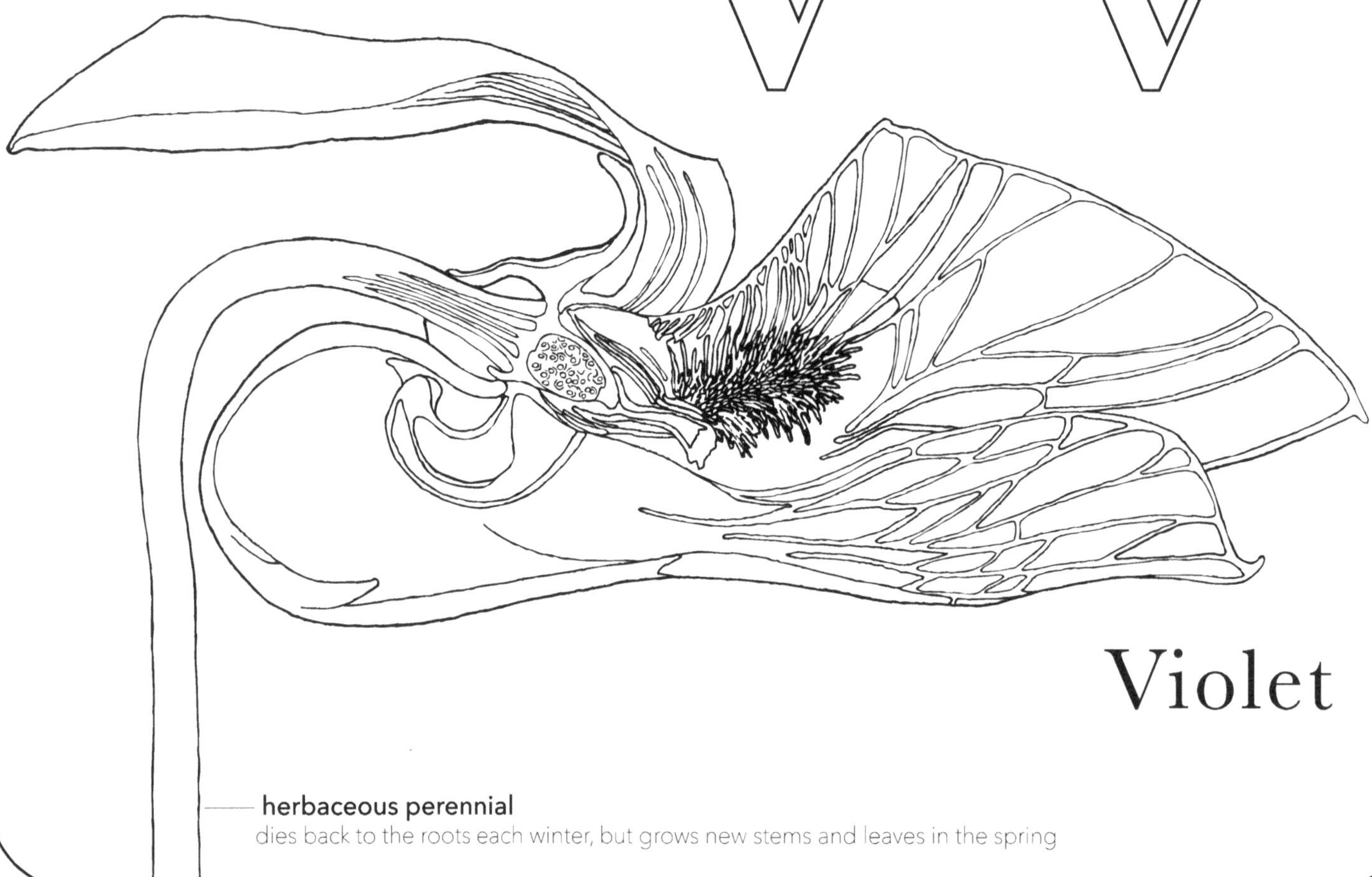

Violet

herbaceous perennial

dies back to the roots each winter, but grows new stems and leaves in the spring

woody perennial
Woody perennials can be deciduous, like wisteria, or evergreen, like the gardenia.

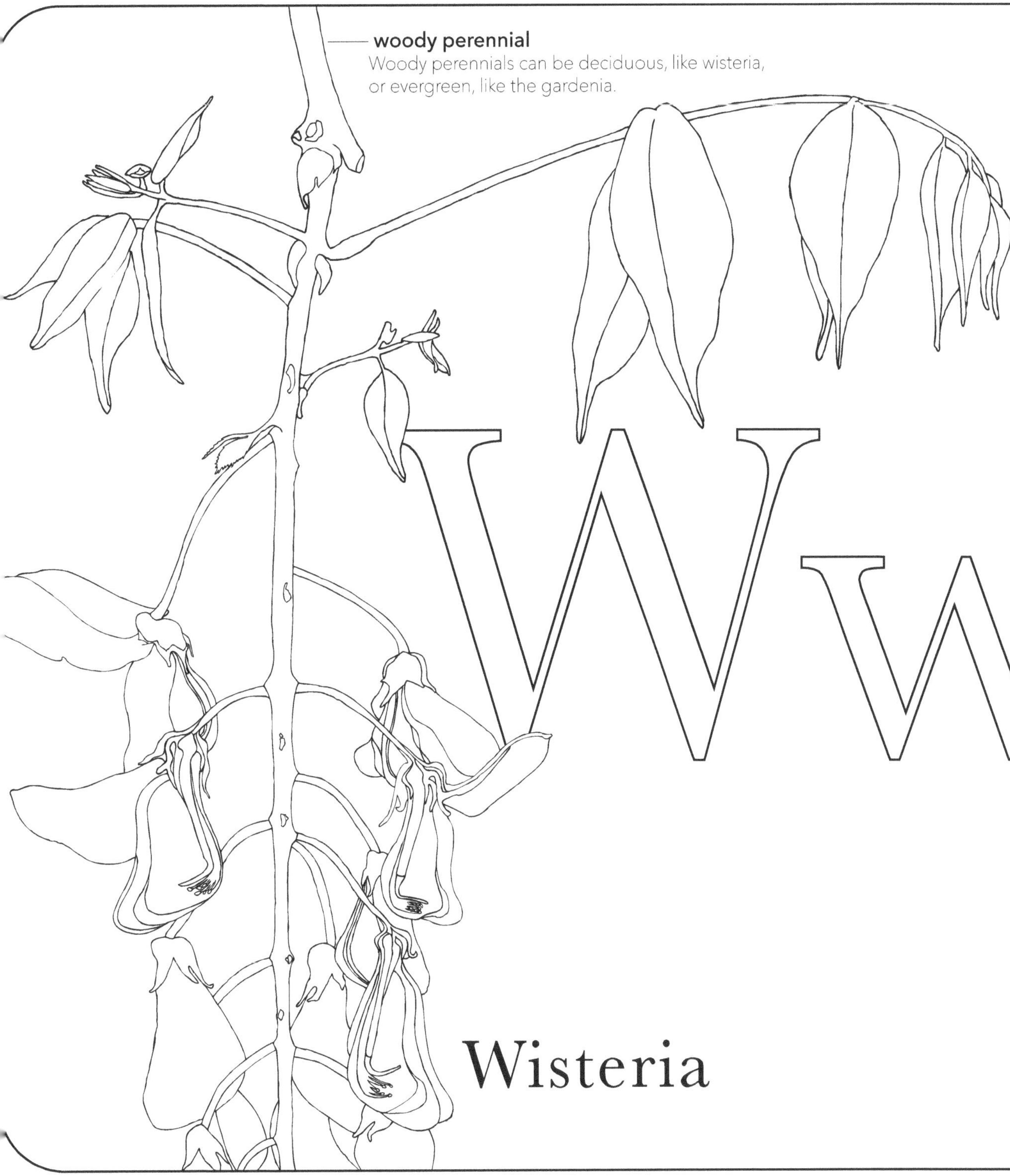

Ww

Wisteria

Xx

tubular flower shape
Foxglove is perfect for bumblebees.
guard hairs
prevent small insects
from entering

foXglove

Y y

pollinator
An animal that transfers pollen, allowing flowers to produce seeds.

Some flowers, like sunflowers, are visited by many pollinators. Others, like foxglove and yucca, have a specific pollinator. Soapweed yucca (not the one drawn here) is only pollinated by the yucca moth.

Yucca

annuals

Annuals complete their life cycle and die in one year (annus). Nasturtium and sunflower are also annuals, but most of the plants in this book are perennials: they live through (peri-) the years (annus). Some plants can be grown as annuals in some climates and perennials in others.

Zz

Zinnia

0

zero

one spur

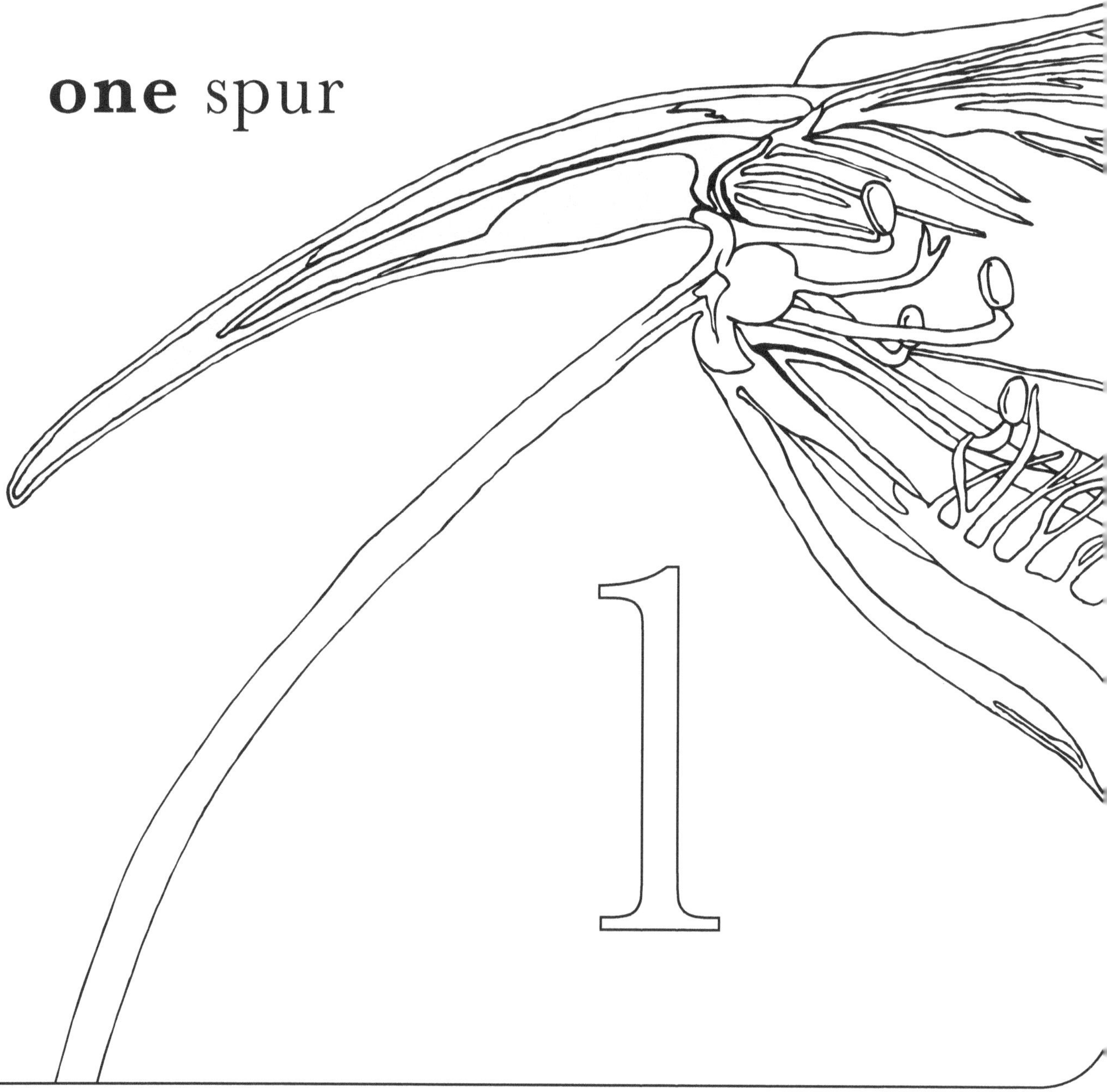

1

2

two thorns

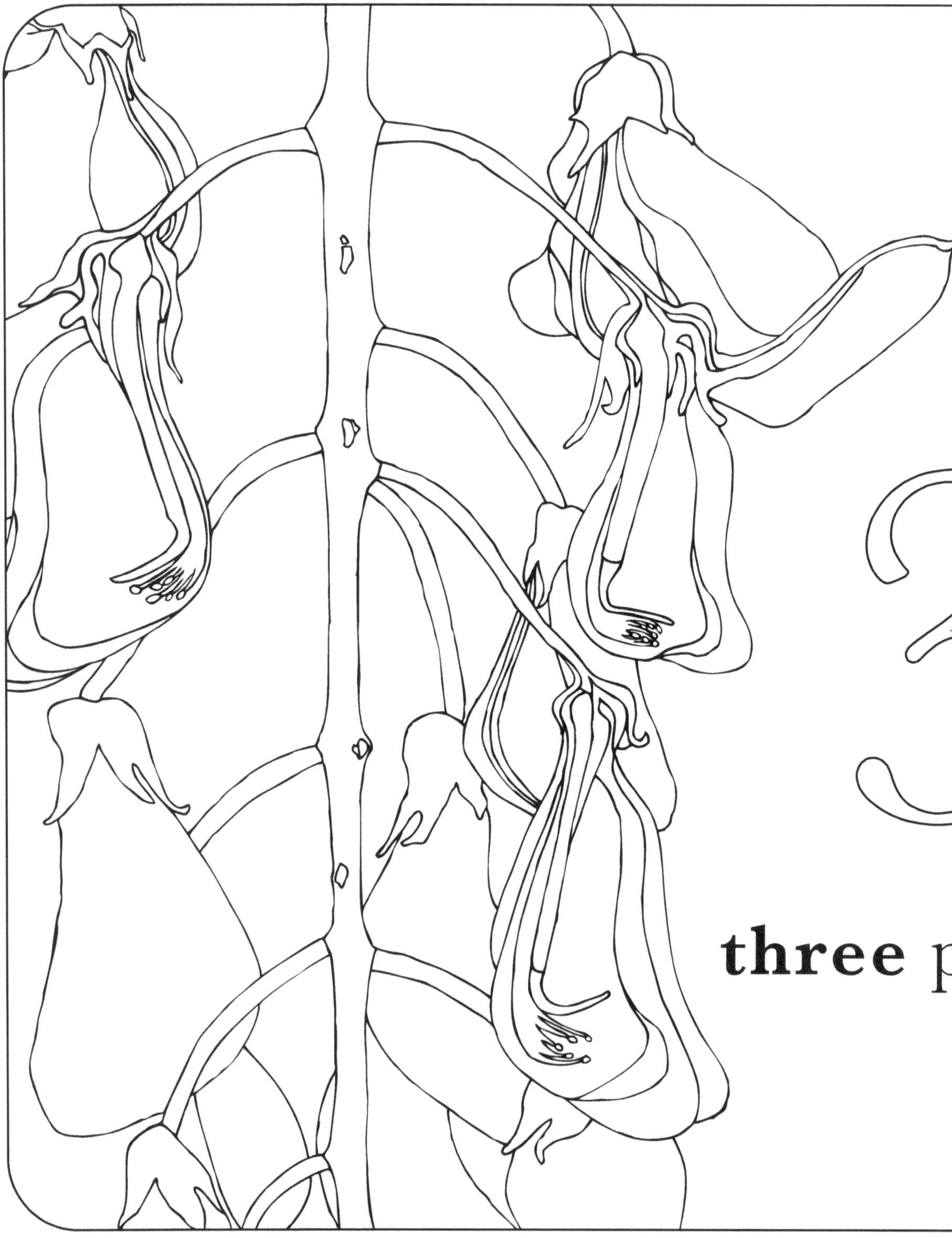

3

three pistils

four stamens

4

5

five tepals

6

six petals

7

seven sepals

8

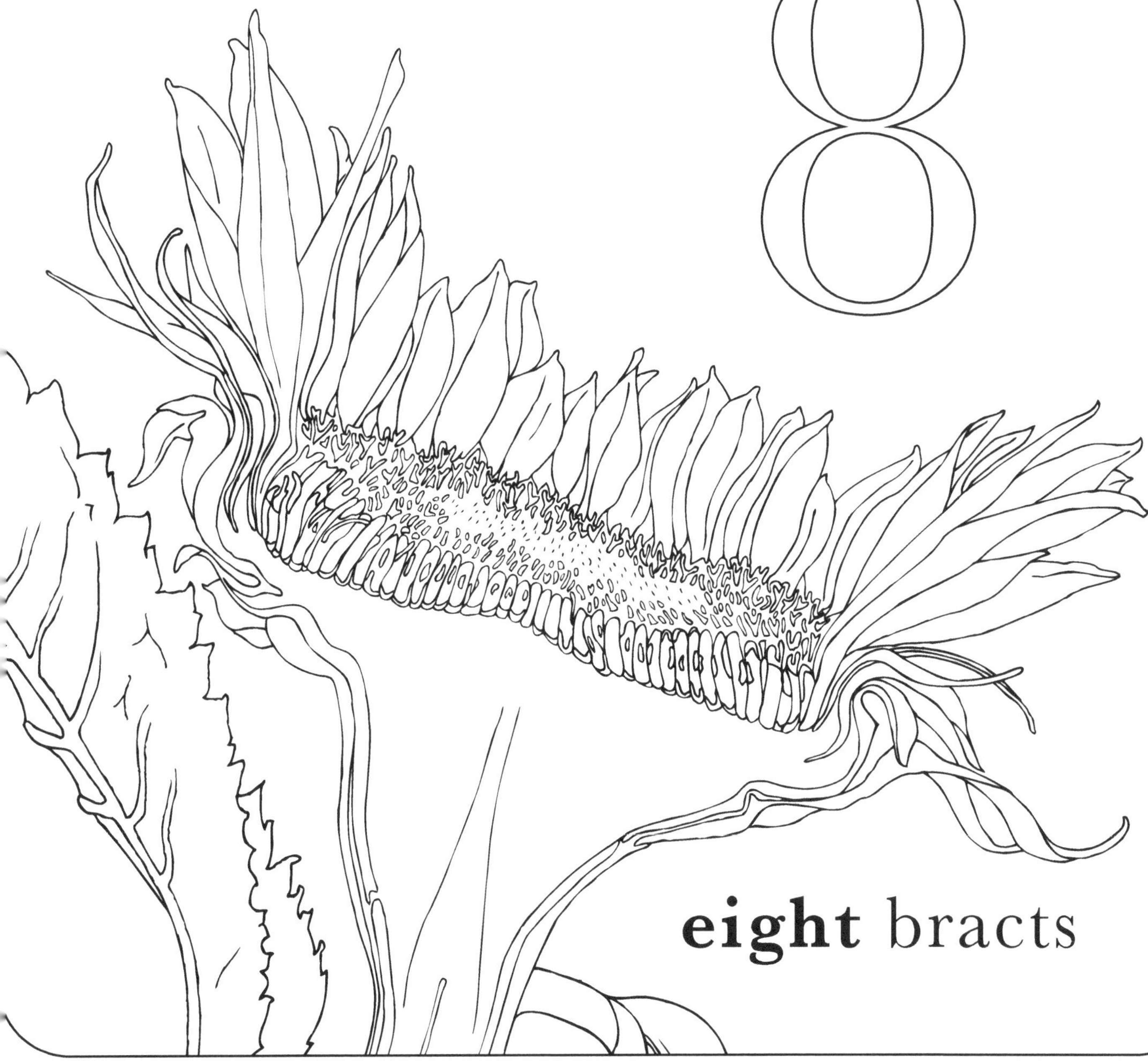

eight bracts

nine nectar guides

9

Botany terms and where to find them in this book:

annuals - plants that complete thier **life cycle** (seed to plant to flower to fruit with ripe seeds) and die in one year. **Zz**

bract - modified leaf at the flower base; can be showy (poinsettia), protective (sunflower), or papery (daffodil). **Dd, 8**

deciduous - plants that lose their leaves in fall and regrow them in the spring. **Ff**

evergreen - plants that have leaves all year long. **Gg**

flower - seed-producing part of a plant that has sepals/tepals and petals/tepals that surround the pistil and/or stamens. Tiny flowers in a capitulum inflorescence are florets. An immature flower is called a bud. **Ss, Tt**

fruit - final stage of the ovary; contains the seeds of the plant. **Pp**

inflorescence - how flowers are arranged on a stem. ie. solitary (single), capitulum (many florets together). **Dd, Ee, Ss**

leaves - flat structures, extending from the stem, where **photosynthesis** (turning sun, water, and carobon dioxide into sugar) takes place; classified by venation (how veins connect), margin (edge), and shape (silhouette). **Qq, Uu**

nectary - most flowers produce nectar and store it in **nectaries** (protected spaces at the base of the flower). **Hh, Nn**

nectar guides - coloration that shows pollinators where to find nectar; may not be visible to humans. **Oo, 9**

ovary - protected area at the base of the pistil where **ovules** become seeds; ovaries become fruits. **Aa, Bb, Hh, Ii, 3**

perianth - the **calyx** (all the sepals of one flower) and **corolla** (all the petals of one flower). **Kk**

perennial - plants that live for many years; may be **herbaceous** (die back underground each year) or woody. **Vv, Ww**

petal - showy flower part that attracts pollinators; often decorated with nectar guides. **Jj, 6**

pistil - female part of a flower; made of the **stigma/s** (sticky tip), **style/s** (long, thin connection), and ovary. **Aa, Cc, 3**

pith - soft, spongy tissue that stores nutrients. **Ee**

plant - multi-celled living organism that uses photosynthesis to create packets of energy (sugar). **Aa-Zz, 0-9**

pollinators - animals that transfer pollen from the anther to the stigma, allowing the flower to produce seeds. Flowers can attract or repel pollinators with specific defenses (ie. guard hairs) and shapes (ie. tubular). **Xx, Yy**

prickle - sharp points that emerge all along the stem but are not connected to the vascular system (veins). **Rr**

seed- a tiny future plant and the nutrition it needs to begin development; enclosed in a protective seed coat (**testa**).

sepal - layer of petal-like (hellebore) or leaf-like (jasmine) parts that protect the petals as they develop. **Hh, Jj, 7**

stamen - male part of a flower; made of an **anther** (where pollen forms) and **filament** (attaches it to the flower). **Aa, 4**

stem - central part of the plant that provides structure; often used to grow more plants without needing seeds. **Cc**

tepals - petals and sepals that are look nearly identical; commonly misidentified as petals. **Ii, Ll, Mm, Tt, 5**

one spur
1
two spines
2
3
three pistils
four stamina
4
5
five tepals
Qq
Quince
Ww
Wisteria
Echinacea
Ee
Rose
Rr
Tt
Aa
Azalea
Sunflower
Ss
Daffodil
Dd
Ff
Forsythia
Zz
Zinnia
Xx
foXglove
Christmas Cactus
Cc

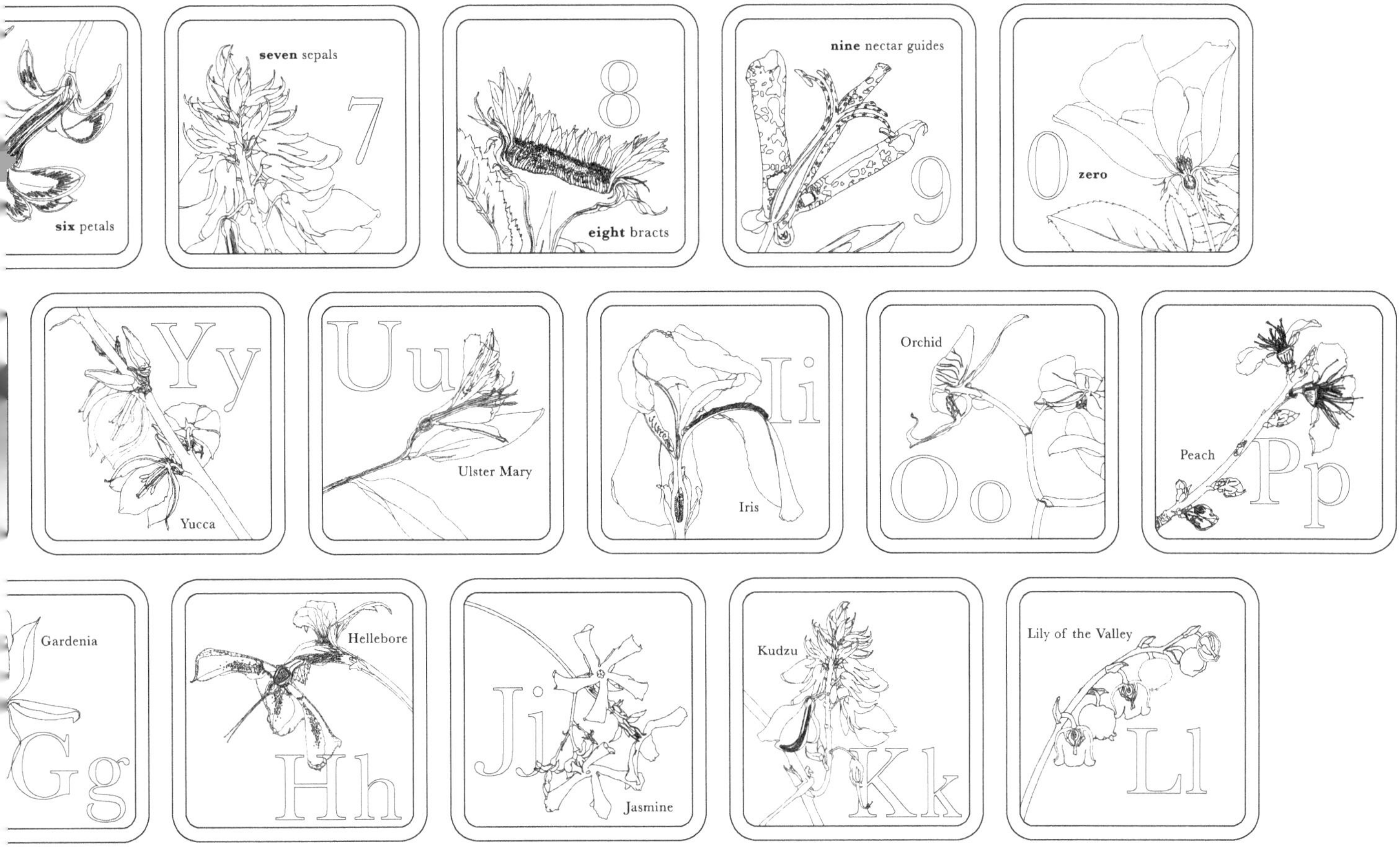

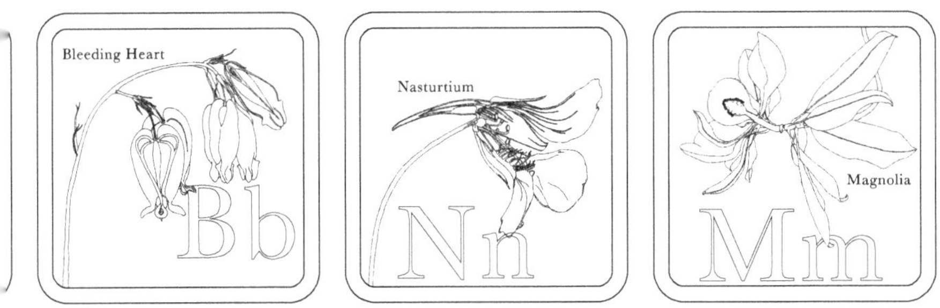

© 2026 Eileen Chevalier
all rights reserved.

www.ingramcontent.com/pod-product-compliance
Lightning Source LLC
Chambersburg PA
CBHW041749220926
51847CB00013B/148
9781952704147